小荧星艺术学校通用教程 / General Tutorial Series of Little Star Art School

模特初级教程
Tutorial for Model Beginner

沈 莹 主编

Editor-in-chief: Shen Ying

上海科学技术文献出版社

Shanghai Scientific and Technological Literature Press

图书在版编目（CIP）数据

模特初级教程 / 沈莹主编 . —上海：上海科学技术文献出版社，2018
小荧星艺术学校通用教程
ISBN 978-7-5439-7678-8

Ⅰ.①模… Ⅱ.①沈… Ⅲ.①时装模特—表演艺术—艺术学校—教材 Ⅳ.①TS942.2

中国版本图书馆CIP数据核字(2018)第213497号

责任编辑：梅雪林　李　峰
封面设计：何子威　李　峰
内文插图：姚　洁
英文翻译：陈路云

模特初级教程
MOTE CHUJI JIAOCHENG
沈　莹　主编
出版发行：上海科学技术文献出版社
地　　址：上海市长乐路746号
邮政编码：200040
经　　销：全国新华书店
印　　刷：常熟市文化印刷有限公司
开　　本：787×1092　1/12
印　　张：11 1/3
版　　次：2019年1月第1版　2019年1月第1次印刷
书　　号：ISBN 978-7-5439-7678-8
定　　价：58.00元
http://www.sstlp.com

序 preface

从 1974 年成立兴趣小组，1985 年"小荧星"品牌创立，实际上，小荧星致力于少儿艺术教育已有四十多年，褪去青涩，陪伴一代又一代的上海小囡成长。蓦然回首，这个艺术的摇篮，已经培育出一批又一批的青年才俊，他们活跃在世界各地的各行各业。

從 1974 年成立興趣小組，1985 年"小熒星"品牌創立，實際上，小熒星致力於少兒藝術教育已有四十多年，褪去青澀，陪伴一代又一代的上海小囡成長。驀然回首，這個藝術的搖籃，已經培育出一批又一批的青年才俊，他們活躍在世界各地的各行各業。

From the time of founding the folk dance interest group in 1974, and the establishment of the brand "Little Star" in 1985, as a matter of fact, Little Star has already devoted to artistic education of children for more than forty years. As its immaturity casting off generally, Little Star has witnessed the grown-up of little girls in Shanghai from one generation to another. When we look back today, groups of young talents have been educated in this cradle of art and stay active in various walks of life.

把孩子培养成富有艺术修养、充满创造活力的时代人才，是小荧星一贯秉承的教育宗旨。"小荧星"品牌创立以来三十余载的辛勤耕耘，老师们认识到，只有让孩子们在学习艺术的过程中真正体验快乐，才能收获自信，从而激发他们的内在潜质。艺术的人生才是完整的人生，从小学会体验艺术，学会发现美，才会丰富人的心灵，从而提升一个人的修养。艺术教育的最终目的不是为了培养多少艺术从业人员，而是提高人的素养。要做到这点，少儿艺术教育必须根据少年儿童的心理与生理特点，并且将系统的、规范的教学理念，通过循序渐进的方法来指导他们的艺术活动，这样才能将少儿趣味与艺术表现融为一体，使理论与实践一致、目标与成果统一。

　　把孩子培養成富有藝術修養、充滿創造活力的時代人才，是小熒星一貫秉承的教育宗旨。"小熒星"品牌創立以來三十餘載的辛勤耕耘，老師們認識到，只有讓孩子們在學習藝術的過程中真正體驗快樂，才能收穫自信，從而激發他們的內在潛質。藝術的人生才是完整的人生，從小學會體驗藝術，學會發現美，才會豐富人的心靈，從而提升一個人的修養。藝術教育的最終目的不是爲了培養多少藝術從業人員，而是提高人的素養。要做到這點，少兒藝術教育必須根據少年兒童的心理與生理特點，并且將系統的、規範的教學理念，通過循序漸進的方法來指導他們的藝術活動，這樣才能將少兒趣味與藝術表現融爲一體，使理論與實踐一致、目標與成果統一。

　　To train our children into talents of the new age who are rich in artistic culture and creative energy is the all along educational purpose of Little Star. Since the brand was founded, our teachers have realized after over thirty years' hard work that the only way for children to gain enough self-confidence and then to inspire their inner potential is to make them really happy during the learning process of art. The life with art is the life with consummation, and one can only have a rich soul and with enhanced accomplishment after he/she masters to experience the art and to discover the beauty in life. To reach that point, the artistic education of children shall be conducted in line with the mental and physical features of children, and shall guide their artistic activities with systematic and normative teaching principles through step-by-step methods. In that case, the preference of children and artistic expression could be united and then to make the theory and practice, or the intention and result centralized.

  正是这个愿望，小荧星团队的全体优秀教师、凝聚全社会艺术精英的小荧星艺委会全体成员，根据多年的教学积累，提炼、汇集教学中最为经典、最具训练价值的内容，编写了这套适用于少年儿童艺术基础训练和各种演出需要的读物，即"小荧星艺术学校通用教程"。这套教程集"歌、舞、演"三大类，其中包括了拉丁舞、芭蕾舞、民族舞、歌舞、主持、合唱、影视共七个不同的专业。同时，为方便大家更直观地学习和欣赏，教程中加入了二维码扫描功能，并设置了互动环节，以帮助家长与孩子之间的互动，将学习、游戏、教育融为一体。

  正是這個願望，小熒星團隊的全體優秀教師、凝聚全社會藝術精英的小熒星藝委會全體成員，根據多年的教學積累，提煉、匯集教學中最爲經典、最具訓練價值的內容，編寫了這套適用於少年兒童藝術基礎訓練和各種演出需要的讀物，即"小熒星藝術學校通用教程"。這套教程集"歌、舞、演"三大類，其中包括了拉丁舞、芭蕾舞、民族舞、歌舞、主持、合唱、影視共七個不同的專業。同時，爲方便大家更直觀地學習和欣賞，教程中加入了二維碼掃描功能，並設置了互動環節，以幫助家長與孩子之間的互動，將學習、遊戲、教育融爲一體。

And that is the very aspiration that makes the splendid teaching team of Little Star and the Little Star Arts Council that gather picked heroes of art in our society, with essential, classical, and most valuable contents through their teaching experiences of ages, create this series of reading materials just suitable for children's basic artistic training and for all kinds of needs for performance, namely, the "General Tutorial Series of Little Star Art School". This series of tutorial includes the three major genera of "singing, dancing and acting", which contain totally seven subjects of Latin dance, ballet, folk dance, song and dance performance, hosting, chorus, film and television. Meanwhile, for the intuitive learning purpose, we include the QR code function and interactive sections in our tutorials, so that our readers can integrate learning, gaming and educating in their parent-child interaction through reading process.

艺术教育是点亮孩子们心智的火焰。希望这套教程犹如浩瀚夜空中萤火虫的闪闪光亮，照亮孩子们的艺术之路，陪伴着孩子们快乐成长。

藝術教育是點亮孩子們心智的火焰。希望這套教程猶如浩瀚夜空中螢火蟲的閃閃光亮，照亮孩子們的藝術之路，陪伴著孩子們快樂成長。

The artistic education is a flame that lights the intelligence in our children's hearts. We truly expect that this series of tutorial could light up the road to art for children like fireflies glittering in the dark night, and could accompany the children to grow up happily.

2016.6

编辑委员会

主　　编： 沈　莹

统　　筹： 陈　斌　杨方禹

执行主编： 易爱博　孟祥麟

编　　委：（按姓氏笔画排名）

　　　　　　王　鹏　王　璨　水　芳　叶　姬　朱诗玥　刘亚雯

　　　　　　吴祎昊　邹树金　柏擎阳　施　琼　魏　蕾

指　　导：（按姓氏笔画排名）

　　　　　　万纪敏　马文婷　王珊珊　冉鹏程　孙　庹　杨　进

　　　　　　邹　晟　汪　岩　张　凤　张伟民　张家雄　张薇倩

　　　　　　范　莉　范笑旻　季彦栋　郭　俊　戚超平　曹燕芸

　　　　　　谢　菲　游家意　蔡来艺

本册撰稿： 水　芳　赵怡琳

美　　术： 何子威　徐　沂

編輯委員會

主　　編：沈　瑩

統　　籌：陳　斌　楊方禹

執行主編：易愛博　孟祥麟

編　　委：（按姓氏簡體筆畫排名）

　　　　　王　鵬　王　璨　水　芳　葉　姬　朱詩玥　劉亞雯

　　　　　吳禕昊　鄒樹金　柏擎陽　施　瓊　魏　蕾

指　　導：（按姓氏簡體筆畫排名）

　　　　　萬紀敏　馬文婷　王珊珊　冉鵬程　孫　庹　楊　進

　　　　　鄒　晟　汪　岩　張　鳳　張偉民　張家雄　張薇倩

　　　　　范　莉　范笑旻　季彥棟　郭　俊　戚超平　曹燕蕓

　　　　　謝　菲　游家意　蔡來藝

本冊撰稿：水　芳　趙怡琳

美　　術：何子威　徐　沂

Editor-in-chief: Shen Ying

Coordinator: Chen Bin, Yang Fangyu

Executive Editor-in-chief: Yi Aibo, Meng Xianglin

Editorial Board Member: (listed in the surname stroke order)

    Wang Peng, Wang Can, Shui Fang, Ye Ji, Zhu Shiyue, Liu Yawen, Wu Yihao, Zou Shujin, Bai Qingyang, Shi Qiong, Wei Lei

Director: (listed in the surname stroke order)

    Wan Jimin, Ma Wenting, Wang Shanshan, Ran Pengcheng, Sun Tuo, Yang Jin, Zou Sheng, Wang Yan, Zhang Feng, Zhang Weimin, Zhang Jiaxiong, Zhang Weiqian, Fan Li, Fan Xiaomin, Ji Yandong, Guo Jun, Qi Chaoping, Cao Yanyun, Xie Fei, You Jiayi, Cai Laiyi

Copywriter: Shui Fang, Zhao Yilin

Artistic Designer: He Ziwei, Xu Yi

目录 / 目錄
Contents

1 编者的话
  編者的話
  Words from editor

第一单元 / 第一單元
Unit One

11 1. 热身训练
   熱身訓練
   Warm-up Exercise

12 2. 方位感知训练
   方位感知訓練
   Direction Perception Exercise

14 3. 模特基础训练
   模特基礎訓練
   Basic Model Exercise

第二单元 / 第二單元
Unit Two

23 1. 热身训练
   熱身訓練
   Warm-up Exercise

24 2. 方位感知训练
   方位感知訓練
   Direction Perception Exercise

28 3. 模特基础训练
   模特基礎訓練
   Basic Model Exercise

37 4. 主题秀
   主題秀
   Theme Show

扫一扫本书中的二维码可查看对应的音频、视频内容。/ 掃一掃本書中的二維碼可查看對應的音頻、視頻內容。
You can watch the corresponding audio & video contents by scanning all the QR codes in this book.

第三单元 / 第三單元
Unit Three

43　1. 热身训练
　　　熱身訓練
　　　Warm-up Exercise

46　2. 方位感知训练
　　　方位感知訓練
　　　Direction Perception Exercise

第四单元 / 第四單元
Unit Four

59　1. 热身训练
　　　熱身訓練
　　　Warm-up Exercise

61　2. 方位感知训练
　　　方位感知訓練
　　　Direction Perception Exercise

63　3. 模特基础训练
　　　模特基礎訓練
　　　Basic Model Exercise

67　4. 主题秀
　　　主題秀
　　　Theme Show

第五单元 / 第五單元
Unit Five

73　1. 热身训练
　　　热身訓練
　　　Warm-up Exercise

74　2. 方位感知训练
　　　方位感知訓練
　　　Direction Perception Exercise

78　3. 主题秀
　　　主題秀
　　　Theme Show

第六单元 / 第六單元
Unit Six

85　1. 热身训练
　　　热身訓練
　　　Warm-up Exercise

86　2. 方位感知训练
　　　方位感知訓練
　　　Direction Perception Exercise

87　3. 模特基础训练
　　　模特基礎訓練
　　　Basic Model Exercise

90　4. 主题秀
　　　主題秀
　　　Theme Show

第七单元 / 第七單元
Unit Seven

97   1. 热身训练
      熱身訓練
      Warm-up exercise

98   2. 方位感知训练
      方位感知訓練
      Direction Perception Exercise

99   3. 模特基础训练
      模特基礎訓練
      Basic Model Exercise

101   4. 主题秀
      主題秀
      Theme Show

第八单元 / 第八單元
Unit Eight

107   1. 模特展示技巧
      模特展示技巧
      Demonstration Skills for Models

111   2. 日常礼仪
      日常禮儀
      Daily Etiquette

116   后记
      後記
      Postscript

118   星星墙
      星星牆
      Wall of Little Stars

编者的话 / 編者的話
Words from editor

　　模特,最早源于英语"Model"的音译,指在体型、样貌、气质、文化基础、职业素养、展示能力等方面具有一定条件,且具备时尚领悟能力,并可在T台、平面、广告中自信展示的表演者。而少儿模特更是美的传播者,能够代表广大少年儿童阳光与积极向上的精神。

　　模特,最早源于英語"Model"的音譯,指在體型、樣貌、氣質、文化基礎、職業素養、展示能力等方面具有一定條件,且具備時尚領悟能力,并可在T台、平面、廣告中自信展示的表演者。而少兒模特更是美的傳播者,能够代表廣大少年兒童陽光與積極向上的精神。

　　The Chinese term 模特 derived from the pronunciation of the English term "Model", which refers to the performers who meet certain standards in terms of body shape, appearance, temperament, knowledge, professional quality, self-displaying ability. These performers can comprehend fashion well, and display themselves confidently on catwalk and in advertisements. As for teenager models, they can better demonstrate beaut, and represent the optimism of children.

　　在少儿模特训练中,通过纠正儿童的不良体态,改善协调性,激发表演欲望,提高对音乐的感悟及时尚的审美能力,对儿童的个性与气质有着积极的推动作用,同时树立儿童的自信,提升个人气场,促进身心健康。

　　在少兒模特訓練中,通過糾正兒童的不良體態,改善協調性,激發表演欲望,提高對音樂的感悟及時尚的審美能力,對兒童的個性與氣質有着積極的推動作用,同時樹立兒童的自信,提升個人氣場,促進身心健康。

　　The training of teenager models will improve their posture and coordination, stimulate their performing desire, enhance their comprehension ability of music and fashion, improve their personality and temperament as well as build confidence, promote charisma and improve health.

《模特初级教程》是"小荧星艺术学校通用教程"之一,是针对四岁以后儿童编写的一部模特教材。全书共有八单元,每个单元均设"要做到""我能行""小提示""蒲公英"。其中"要做到"板块主要从热身、方位感知、模特基础着手训练,培养儿童优美挺拔的身姿与多种行走转身展示技法的学习。"我能行"板块从儿童的视角对动作示范进行图示讲解。"小提示"与"蒲公英"板块则是对教学要点,组合做法等内容进行分析,帮助儿童掌握技术技巧。

《模特初級教程》是"小熒星藝術學校通用教程"之一,是針對四歲以後兒童編寫的一部模特教材。全書共有八單元,每個單元均設"要做到""我能行""小提示""蒲公英"。其中"要做到"板塊主要從熱身、方位感知、模特基礎著手訓練,培養兒童優美挺拔的身姿與多種行走轉身展示技法的學習。"我能行"板塊從兒童的視角對動作示範進行圖示講解。"小提示"與"蒲公英"板塊則是對教學要點,組合做法等內容進行分析,幫助兒童掌握技術技巧。

Elementary Model Textbook is one of the general art textbooks of Shanghai Little Star. This model textbook is written for kids over 4 years old. There are 8 units in the textbook. Each unit includes sections such as "Key Points", "I Can Do It", "Notes" and "Dandelion". "Key Points" focuses on warm-up exercise, direction perception exercise and basic model exercise, improving kids' posture and their walking and turning skills. "I Can Do It" mainly explains the movements with pictures from kids' perspective. "Notes" and "Dandelion" mainly analyzes the key points and some combinations, which enables kids to acquire skills.

主要教学目的 / 主要教學目的
Main teaching objectives

1. 激发儿童学习兴趣,对少儿模特有初步的认知。

   激發兒童學習興趣,對少兒模特有初步的認知。

   Stimulate kids' interest in learning, and give them a glimpse of teenager models.

2. 纠正不良体态，塑造挺拔身姿。

　　纠正不良體態，塑造挺拔身姿。

　　Improve kids' body shape and posture.

3. 引导儿童对服装色彩及风格有初步的审美认知，学习基础平面拍摄、基础定位及路线行走。

　　引導兒童對服裝色彩及風格有初步的審美認知，學習基礎平面拍攝、基礎定位及路線行走。

　　Guide kids to have a preliminary aesthetic understanding of clothing color and style. Teach them basic shooting skills, basic positioning and walking skills.

4. 树立自信，提升情感领悟能力。

　　樹立自信，提升情感領悟能力。

　　Help kids build confidence and enhance their comprehension ability.

给家长的建议 / 給家長的建議
Advice to parents

少儿模特是体现儿童个性，展示美的艺术，它传播着健康与积极向上的精神，为此，我们建议家长：

少兒模特是體現兒童個性，展示美的藝術，它傳播着健康與積極向上的精神，爲此，我們建議家長：

Teenager modeling is an art form which demonstrates children's personality, displays beauty and spreads a healthy and positive spirit. As a result, we'd like to give parents some advice as follows:

1. 要爱护孩子，不要急功近利。儿童的心理和生理是很稚嫩的，选择合适的舞台，不要给他们太大的压力，切忌拔苗助长。

　　要愛護孩子，不要急功近利。兒童的心理和生理是很稚嫩的，選擇合適的舞台，不要給他們太大的壓力，切忌拔苗助長。

Love your kids and do not hurry. Children are psychologically and biologically weak. As parents, you should make appropriate decisions. Do not give them too much pressure.

2. 要多鼓励孩子，让他们建立起自信。模特需要绝对的自信，靠自身散发的独特气场来展示服装，吸引观众，一旦自信心丧失了，这一切也就失去了依靠。

要多鼓勵孩子，讓他們建立起自信。模特需要絕對的自信，靠自身散發的獨特氣場來展示服裝，吸引觀眾，一旦自信心喪失了，這一切也就失去了依靠。

Encourage your kids so as to help them build confidence. Models should be absolutely confident. They display clothes and attract audience by their own charisma. If they lose confidence, they can not achieve their goals as models.

3. 要为孩子们营造一个良好的学习环境，让孩子们受到良好的熏陶，这会潜移默化帮助他们的成长，这样才能让孩子表现出符合自身年龄的特质与气场。

要爲孩子們營造一個良好的學習環境，讓孩子們受到良好的熏陶，這會潛移默化幫助他們的成長，這樣才能讓孩子表現出符合自身年齡的特質與氣場。

Create a good learning environment for children. A good environment will profoundly and positively influence their growth. Only in this way can kids demonstrate the characteristics and charisma of their own age.

《小荧星艺术学校通用教程》编写组

《小熒星藝術學校通用教程》編寫組

Writing Group of General Textbooks of Shanghai Little Star Art School

2018.4

第一单元

如果你认为学模特只是学走路而已，

那谁不会走路呢？

为什么同样是走路超模们却自带光环？

如何成为秀场上的焦点，

如何用眼神 HOLD 住全场，这是一门学问。

想成为一名时尚潮童，必须学会很多知识与技能。

本单元我们就从基础站姿开始练习吧。

第 一 單 元

如果你認為學模特祇是學走路而已，

那誰不會走路呢？

為什麼同樣是走路超模們卻自帶光環？

如何成為秀場上的焦點，

如何用眼神 HOLD 住全場，這是一門學問。

想成為一名時尚潮童，必須學會很多知識與技能。

本單元我們就從基礎站姿開始練習吧。

Unit One

If you think that learning to become a model is just learning to walk, then who on earth can't do that? However, why do those supermodels look so attractive when they are walking? How to become the focus of a show and use eye contact to hold all the audience is something worth learning. If you want to become fashionable, you must acquire a lot of knowledge and skills. In this unit, let's start with the basic practices of stance.

1. 热身训练 / 熱身訓練
Warm-up Exercise

认识身体 / 認識身體
Know your body

要做到 / 要做到
Key Points

通过头、肩、腰等部位的活动，使身体各部位具有灵活性。

通過頭、肩、腰等部位的活動，使身體各部位具有靈活性。

Develop the flexibility of different parts of your body through activities involving head, shoulders and waist, etc.

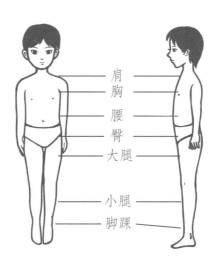

2. 方位感知训练 / 方位感知訓練
Direction Perception Exercise

要做到 / 要做到
Key Points

认识教室八个方位。

認識教室八個方位。

Understand the eight directions in classroom.

1 点—正前方

1 點—正前方

Point 1—front

2 点—右斜前方

2 點—右斜前方

Point 2—right front

3 点—右方

3 點—右方

Point 3—right

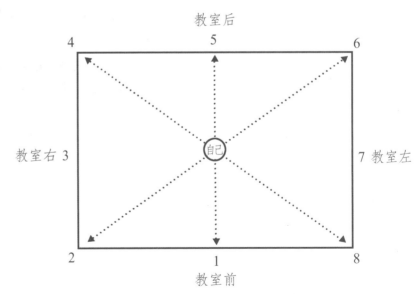

4 点—右斜后方

4 點—右斜後方

Point 4—right rear

5 点—后方

5 點—後方

Point 5—rear

6 点—左斜后方

6 點—左斜後方

Point 6—left rear

7 点—左方

7 點—左方

Point 7—left

8 点—左斜前方

8 點—左斜前方

Point 8—left front

小组合 / 小組合
Small Motion Combinations

《找方位》

《找方位》

"find directions"

3. 模特基础训练 / 模特基礎訓練
Basic Model Exercise

(1) **基本站姿** / 基本站姿
Basic Stance

要做到 / 要做到
Key Points

　　头摆正、双眼平视、肩颈放松、挺胸、收腹、立腰、直背、收臀、直膝。

　　頭擺正、雙眼平視、肩頸放鬆、挺胸、收腹、立腰、直背、收臀、直膝。

　　You should keep your head upright, look straight ahead, relax your neck and shoulders and straighten up.

小提示 / 小提示
Notes

　　脚—正步位、脚尖超1点方向。
　　腳—正步位、腳尖超1點方向。
　　Feet—Positive step: keep tiptoes facing the front.

小腿—小腿内侧肌加紧。

小腿—小腿內側肌加緊。

Calves—tighten medial calf muscles.

膝—膝关节伸直,内侧靠拢。

膝—膝關節伸直,內側靠攏。

Knees—straighten knee joints and stand with knees together.

大腿—大腿内侧肌加紧。

大腿—大腿內側肌加緊。

Thighs—tighten medial thigh muscles.

髋—髋骨摆正,臀大肌由外向里收紧。

髖—髖骨擺正,臀大肌由外向裡收緊。

Hips—straighten hips and tighten gluteus maximus.

腹—收紧腹肌。

腹—收緊腹肌。

Belly—tighten abdominal muscles.

胸—胸部挺起,肩胛骨打开。

胸—胸部挺起,肩胛骨打開。

Chest—keep the chest forward and shoulders back.

颈—颈椎关节直立,有拉长的感觉。

頸—頸椎關節直立,有拉長的感覺。

Neck—keep the neck erect and feel it is extended.

头—额头有向上顶的感觉，目视前方。

頭—額頭有向上頂的感覺，目視前方。

Head—feel the forehead seems to move upward and look forward.

我能行 / 我能行
I Can Do It

双手垂于身体两侧站立　　　双手叉腰站　　　单手叉腰站

蒲公英 / 蒲公英
Dandelion

女生叉腰时，手腕向下压会显得上身更优雅挺拔。

女生叉腰時，手腕向下壓會顯得上身更優雅挺拔。

When girls stand with their hands on the hips, their upper body would seem to be more elegant if they keep the wrists down.

(2) 直线目标行走训练 / 直綫目標行走訓練
Straight Walking Training

要做到 / 要做到
Key Points

从后方踩线，笔直走到前方。

從後方踩綫，筆直走到前方。

walk from the rear line to the front.

对准标志物，大胆向前直线行走。

對準標志物，大膽向前直綫行走。

Aim at the target and walk straight forward boldly.

17

星星班长的话
星星班長的話
Xingxing's Words

第二单元

知道模特吸引观众的秘诀之一是什么吗?

答案是:明亮的眼睛和恰当的表情与造型。

眼睛是心灵的窗户,

恰当的表情与造型能反应出模特对展示产品的理解。

所以,本单元就从眼神及表情入手,

进行模特基础训练,一起来玩儿一下吧。

第 二 單 元

知道模特吸引觀眾的秘訣之一是什麼嗎?

答案是:明亮的眼睛和恰當的表情與造型。

眼睛是心靈的窗戶,

恰當的表情與造型能反應出模特對展示產品的理解。

所以,本單元就從眼神及表情入手,

進行模特基礎訓練,一起來玩兒一下吧。

Unit Two

What is one of the secrets of models to attract audiences?
The answer is: bright eyes, proper expressions and poses.
Since eyes can reflect one's soul,
proper expressions and poses will show models'
understanding of the displayed products.
Therefore, this unit will focus on eyes and facial expressions,
providing you with some basic model training. Let's start.

1. 热身训练 / 熱身訓練
Warm-up Exercise

要做到 / 要做到
Key Points

通过头、肩、腰等部位活动，使身体具有灵活性。

通過頭、肩、腰等部位活動，使身體具有靈活性。

strengthen your body's flexibility with exercises involving head, shoulder and waist, etc.

通过手脚在动作中的协调配合，使身体具有协调性。

通過手腳在動作中的協調配合，使身體具有協調性。

strengthen your body's coordination with the coordinated exercises involving hands and feet.

通过弹跳和伸展使肌肉得到拉伸，增强活力，增长身高，锻炼心肺功能。

通過彈跳和伸展使肌肉得到拉伸，增強活力，增長身高，鍛煉心肺功能。

stretch your muscles by jumping and stretching, which will improve your energy level, height, and cardiopulmonary function.

2. 方位感知训练 / 方位感知訓練
Direction Perception Exercise

眼神训练 / 眼神訓練
Eye Contact Training

要做到 / 要做到
Key Points

> 眼神具有目标感、明亮感。
>
> 眼神具有目標感、明亮感。
>
> have a sense of direction and brightness in your eyes.

我能行 / 我能行
I Can Do It

向上看

向下看

向左看

向左上看

向左下看

向右看

向右上看

向右下看

向正中看

小提示 / 小提示
Notes

除正中外，其余方向要适度，眼珠转的太快会露出大片白眼球，影响美观。

除正中外，其餘方向要適度，眼珠轉的太快會露出大片白眼球，影響美觀。

Despite the middle, you should look in other directions properly. It might look unpleasant if you roll your eyes too quick and a large part of the white is seen by the audiences.

蒲公英 / 蒲公英
Dandelion

台前造型时，眼球转的方向不可超过两个，并且要与头部运动相结合，即头动眼停，眼动头停。

台前造型時，眼球轉的方向不可超過兩個，并且要與頭部運動相結合，即頭動眼停，眼動頭停。

When you pose on the stage, your eyes should roll in no more than two directions. What's more, the rolling of eyes should be combined with the turning of head. When you are turning your head, do not roll your eyes; when you are rolling your eyes, do not turn your head.

小组合 / 小組合
Small Motion Combinations

《明亮的大眼睛》

《明亮的大眼睛》

"Big Bright Eyes"

3. 模特基础训练 / 模特基礎訓練
Basic Model Exercise

(1) 站姿 / 站姿
Stance

要做到 / 要做到
Key Points

肩膀、双脚用力向下延伸。头顶耳根用力向上延伸。

肩膀、雙脚用力向下延伸。頭頂耳根用力向上延伸。

forcibly stretch your shoulders and feet downward, and your head and ears upward.

动作时,均匀呼吸,眼睛平视前方。

動作時,均匀呼吸,眼睛平視前方。

breathe evenly and look straight ahead while doing the above exercises.

双脚平行站姿:

雙脚平行站姿:

Standing with parallel feet:

双脚平行站姿

小提示 / 小提示
Notes

双脚打开，与肩同宽。重心在双脚中间，双手叉腰，身体保持直立。

雙脚打開，與肩同寬。重心在雙脚中間，雙手叉腰，身體保持直立。

Your feet should be apart in shoulder-width, with weight in the middle of the feet. Put your hands on the waist and stand upright.

前后脚站姿：

前後脚站姿：

Standing with unparallel feet

前后脚站姿

小提示 / 小提示
Notes

左脚在前，右脚在后。重心在双脚中间，身体保持直立。

左脚在前，右脚在後。重心在雙脚中間，身體保持直立。

Put your left foot in front of your right foot. Place your weight in the middle of the feet and stand upright.

踮脚站姿：

踮脚站姿：

Standing on tiptoe

小提示 / 小提示
Notes

脚尖踮起，脚跟离地，30度、60度，使重心在脚掌上反复练习。

脚尖踮起，脚跟離地，30度、60度，使重心在脚掌上反複練習。

Stand on tiptoe with your heels off the ground in 30 degrees and 60 degrees. Shift your weight repeatedly on the sole.

踮脚站姿

踮脚站姿

微笑

萌萌哒

(2) 表情训练 / 表情訓練
Training of facial expressions

要做到 / 要做到
Key Points

面部肌肉放松，尝试各种有趣的表情。

面部肌肉放鬆，嘗試各種有趣的表情。

Relax your facial muscles and try on various kinds of interesting facial expressions.

我能行 / 我能行
I Can Do It

惊讶

俏皮

酷

小组合 / 小組合
Small Motion Combinations

《有趣的脸》

《有趣的臉》

"An Interesting Face"

蒲公英 / 蒲公英
Dandelion

好的小模特就像演员一样，通过改变面部表情来表达自己对所展示产品的理解。小模特们快对着镜子练习起来吧！

好的小模特就像演員一樣，通過改變面部表情來表達自己對所展示產品的理解。小模特們快對著鏡子練習起來吧！

Good models are just like actors. They can show their understanding of the displayed products by changing their facial expressions. Let's do some exercises in front of the mirrors!

(3) 台前定位 / 台前定位
Positioning on the stage

a. 正步位单腿屈膝

a. 正步位單腿屈膝

a. Keep your body facing straight ahead with one leg bent

要做到 / 要做到
Key Points

双腿并拢，脚尖朝 1 点方向，单腿屈膝，重心在中间。

雙腿併攏，脚尖朝 1 點方向，單腿屈膝，重心在中間。

Stand with legs together. Your tiptoe should be towards the 1 o'clock direction. With one leg bent, your should put your weight in the middle.

蒲公英 / 蒲公英
Dandelion

小朋友们单腿屈膝时不要出胯哦，这样会使你变矮的。

小朋友們單腿屈膝時不要出胯哦，這樣會使你變矮的。

Don't move your hip when you bent your leg, otherwise you might look shorter.

萤火虫 / 螢火蟲
Firefly

哪个动作是不对的?

哪個動作是不對的?

Which one is incorrect?

模特初级教程 / 模特初级教程
Tutorial for Model Beginner

b. 左右重心的转移

b. 左右重心的轉移

b. Shift weight between left and right

要做到 / 要做到
You should

双脚打开与肩同宽，左脚脚尖朝一点方向，右脚脚尖朝 2 点方向。重心偏左脚。

雙腳打開與肩同寬，左腳腳尖朝一點方向，右腳腳尖朝 2 點方向。重心偏左腳。

Keep your feet apart in shoulder-width, with left tiptoe towards the 1 o'clock direction and right tiptoe towards the 2 o'clock direction. Shift weight to your left foot.

双脚打开与肩同宽，左脚脚尖朝 8 点方向，右脚脚尖朝 1 点方向。重心偏右脚。

雙腳打開與肩同寬，左腳腳尖朝 8 點方向，右腳腳尖朝 1 點方向。重心偏右腳。

Keep your feet apart in shoulder-width, with left tiptoe towards the 8 o'clock direction and right tiptoe towards the 1 o'clock direction. Shift weight to your right foot.

蒲公英 / 蒲公英
Dandelion

男生在做双脚分开造型时，通常将重心放在中间，显得更阳刚一些。

男生在做雙脚分開造型時，通常將重心放在中間，顯得更陽剛一些。

When boys pose with their feet apart, the weight is usually put in the middle, which would make them look more manly.

4. 主题秀 / 主题秀
Theme Show

我能行 / 我能行
I Can Do It

直线行走—台前定位

直綫行走—台前定位

Walk straight forward--Position yourself on the stage

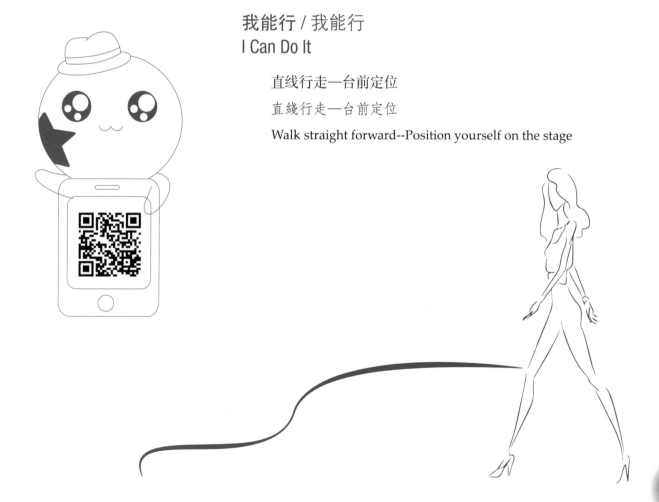

星星班长的话
星星班長的話
Xingxing's Words

第三单元

经过两单元的学习,

你是否对模特已经有一些了解?

是不是更有兴趣了呢?

本单元我们会将学习的内容进行小小的组合,

初步感受一下台前定位的魅力。

来吧,孩子们,充分展现你自己吧!

第三單元

經過兩單元的學習，

你是否對模特已經有一些了解？

是不是更有興趣了呢？

本單元我們會將學習的內容進行小小的組合，

初步感受一下台前定位的魅力。

來吧，孩子們，充分展現你自己吧！

Unit Three

After two units,

have you learned more about models?

Are you more interested about it?

In this unit, we will combine what we've learned in the last two

units to feel the charm of positioning on the stage.

Come on, kids, it is your showtime now!

1. 热身训练 / 熱身訓練
Warm-up Exercise

腰腹练习 / **腰腹練習**
Waist and abdominal training

要做到 / **要做到**
Key Points

双脚与肩同宽站立。

雙腳與肩同寬站立。

Keep your feet apart in shoulder-width.

双臂向两侧平举伸直。

雙臂向兩側平舉伸直。

Stretch your arms sideways.

以肘关节为支点，两手向上弯，大臂与小臂呈90度，掌心相对。

以肘關節爲支點，兩手向上彎，大臂與小臂呈90度，掌心相對。

Raise lowers arms until they are at 90 degrees against upper arms. Keep palms facing each other.

手型不变，身体向左侧拧转，重心偏至左脚。

手型不變，身體向左側撐轉，重心偏至左腳。

Keep hands in the same position, twist your body to the left and shift weight to the left foot.

正步步位预备姿势。

正步步位預備姿勢。

Stand in the positive position for preparation.

身体向左侧弯曲，左手沿大腿外侧向下延伸至最大限度，同时屈右肘。

模特初级教程 / 模特初級教程
Tutorial for Model Beginner

身体向左侧弯曲，左手沿大腿外侧向下延伸至最大限度，同时屈右肘。

Bend your body to the left, extend your left hand to the maximum along the outer thigh, and bend your right arm at the same time.

当左手手掌向下延伸至最大限度时，右手向上伸展。

當左手手掌向下延伸至最大限度時，右手向上伸展。

When the left hand extends down to the maximum, extend your right hand upward.

平板支撑。

平板支撐。

Plank exercise.

蒲公英 / 蒲公英
Dandelion

腰腹部就像连接上下肢的纽带，通过锻炼可以增加韧性和灵活性，改善塌腰等不良体态。

腰腹部就像連接上下肢的紐帶，通過鍛煉可以增加韌性和靈活性，改善塌腰等不良體態。

The waist and abdomen are like a link between the upper and lower limbs. The training can enhance the flexibility of your body, and improve the unhealthy posture such as arc at the back.

2. 方位感知训练 / 方位感知訓練
Direction Perception Exercise

(1) 手型 / 手型
Hand Form

要做到 / 要做到
Key Points

女生——虎口向前，五指伸直姿态中，食指略向上翘，大拇指略往里扣。

女生——虎口向前，五指伸直姿態中，食指略向上翹，大拇指略往裏扣。

Girls——Keep the part between thumb and index forward. Stretch all the fingers, with the index slightly upward and the thumb slightly inward.

男生——虎口向前，五指自然放松呈空心拳状。

男生——虎口向前，五指自然放鬆呈空心拳狀。

Boys——Keep the part between thumb and index forward. Keep the hand naturally clenched.

我能行 / 我能行
I Can Do It

(2) 芭蕾手位 / 芭蕾手位
Ballet Hand Position

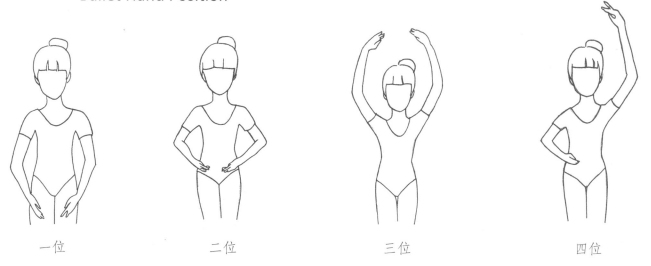

一位　　　　　二位　　　　　三位　　　　　四位

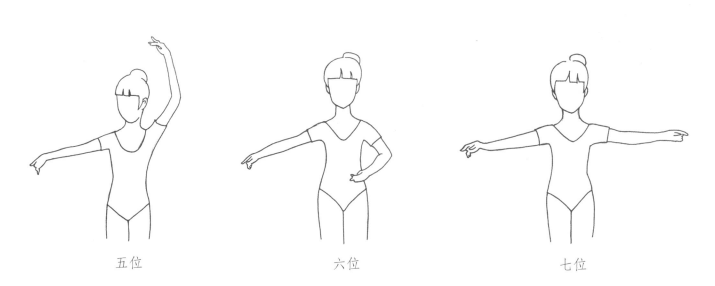

五位　　　　　六位　　　　　七位

小组合 / 小組合
Small Motion Combinations

《春天》

《春天》

"Spring"

(3) 摆臂 / 擺臂
Arm Swing

要做到 / 要做到
Key Points

手臂前后自然摆动，向前约 30 度，向后约 15 度。

手臂前後自然擺動，向前約 30 度，向後約 15 度。

Swing arms naturally. About 30 degrees forward and 15 degrees backward.

大臂带动小臂，小臂带动手，两臂摆动幅度要一致。

大臂帶動小臂，小臂帶動手，兩臂擺動幅度要一致。

Let the upper arm set the lower arm in motion, and let the latter set the hand in motion. Keep the swing amplitude of two arms at same level.

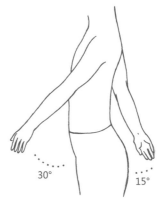

摆臂幅度

(4) 前后重心的转移 / 前後重心的轉移
Forward and Backward Shift of Weight

小组合 / 小組合
Small Motion Combinations

准备：双脚并拢，双手叉腰。

準備：雙脚并攏，雙手叉腰。

Get ready: Stand with your feet together and hands on the waist.

[1]-[4]　左脚自然抬起，朝 1 点方向迈步。

[1]-[4]　左脚自然抬起，朝 1 點方向邁步。

[1]-[4]　Lift your left foot naturally, and move toward Point 1.

　　　　左脚落地，重心转移至左脚，右脚脚尖跐起，膝盖伸直。

　　　　左脚落地，重心轉移至左脚，右脚脚尖跐起，膝蓋伸直。

Shift your weight to the left foot when it reaches the floor. Keep your right foot on tiptoe and right leg straightened.

[5]-[8]　右脚抬起并自然收回。

[5]-[8]　右脚抬起并自然收回。

Lift your right foot and take it back.

　　　　右脚与左脚并拢。

　　　　右脚與左脚併攏。

Keep the right foot together with the left one.

[9]-[12]　右脚自然抬起，朝 1 点方向迈步。

[9]-[12]　右腳自然抬起，朝 1 點方向邁步。

Lift your right foot naturally, and move toward Point 1.

右脚落地，重心转移至右脚，左脚脚尖跷起，膝盖伸直。

右脚落地，重心轉移至右脚，左脚脚尖跷起，膝蓋伸直。

Shift your weight to the right foot when it reaches the floor. Keep your left foot on tiptoe and left leg straightened.

[13]-[16]　左脚抬起并自然收回。

[13]-[16]　左脚抬起并自然收回。

Lift your left foot and take it back.

左脚与右脚并拢。

左脚與右脚併攏。

Keep the left foot together with the right one.

重复 [1]-[16]。

重復 [1]-[16]。

Repeat [1] - [16].

蒲公英 / 蒲公英
Dandelion

小模特们,练好重心的转移是走好模特步的关键之一哦!加油吧!

小模特們,練好重心的轉移是走好模特步的關鍵之一哦!加油吧!

Dear models, learning how to shift weight is one of the keys to walking like professional models. Let's start!

(5) 表情与造型——五指练习法 / 表情與造型——五指練習法
Facial Expressions and Poses—Five-fingers Practice

要做到 / 要做到
Key Points

通过数字的递增记忆，表情与造型相结合，定位准确。

通過數字的遞增記憶，表情與造型相結合，定位準確。

Combine facial expressions and poses through memorizing the increasing of numbers, and position yourself correctly.

我能行 / 我能行
I Can Do It

萤火虫 / 螢火蟲
Firefly

小朋友们，你还能想出哪些可爱的造型呢？

小朋友們，你還能想出哪些可愛的造型呢？

Dear kids, can you think of some other cute poses?

星星班长的话
星星班長的話
Xingxing's Words

第四单元

模特走秀的关键是一个"走"字,
想要"走"好模特步可不是那么简单哦,
如果你有一些走路的坏习惯,赶紧改掉吧!
本单元着重强调如何走好模特步,
如何优雅自然地转身。让我们气场全开,练起来!

第四單元

模特走秀的關鍵是一個"走"字，

想要"走"好模特步可不是那麼簡單哦，

如果你有一些走路的壞習慣，趕緊改掉吧！

本單元着重強調如何走好模特步，

如何優雅自然地轉身。讓我們氣場全開，練起來！

Unit Four

The key to the models' performance on the catwalk is walking well.
However, it is not easy to walk well on the catwalk.
If you have some bad habits in walking,
get rid of them as soon as possible!
This unit mainly talks about how to walk well on the catwalk
and how to turn around gracefully and naturally.
Let's have a try now!

1. 热身训练 / 熱身訓練
Warm-up Exercise

要做到 / 要做到
Key Points

通过单脚跳跃了解左右重心的区别。

通過單腳跳躍了解左右重心的區別。

understand the difference between putting weight on the left and that on the right through jumping with one leg.

通过手、脚与音乐的配合,增加身体的协调能力,使身体关节部位更加灵活。

通過手、腳與音樂的配合,增加身體的協調能力,使身體關節部位更加靈活。

Enhance coordination and flexibility of the body through moving arms and legs with the rhythm of music.

小组合 / 小組合
Small Motion Combinations

参考音乐《哈哈颂》

參考音樂《哈哈頌》

"Ode to Laughter"

2. 方位感知训练 / 方位感知訓練
Direction Perception Exercise

音乐与节奏 / 音樂與節奏
Music and Rhythm

要做到 / 要做到
Key Points

能区分 $\frac{2}{4}$ 拍 $\frac{3}{4}$ 拍 $\frac{4}{4}$ 拍的歌曲，并准确打节奏。

能區分 $\frac{2}{4}$ 拍 $\frac{3}{4}$ 拍 $\frac{4}{4}$ 拍的歌曲，並準確打節奏。

Be able to distinguish among $\frac{2}{4}$ tempo, $\frac{3}{4}$ tempo, $\frac{4}{4}$ tempo and beat time to the music.

小提示 / 小提示
Notes

随着音乐，有节奏的拍手训练。

隨着音樂，有節奏的拍手訓練。

Clap out with the rhythm.

随着音乐，有节奏的踏步训练。

隨著音樂，有節奏的踏步訓練。

Tap with the rhythm.

随着音乐，有节奏的摆臂训练。

隨著音樂，有節奏的擺臂訓練。

Swing arms with the rhythm.

蒲公英 / 蒲公英
Dandelion

一名优秀的小模特，耳朵要会听节奏，肢体能与音乐相结合表现出应有的状态，加油吧孩子们！

一名優秀的小模特，耳朵要會聽節奏，肢體能與音樂相結合表現出應有的狀態，加油吧孩子們！

An excellent model should be good at listening to rhythm and combining your body movement with music. Let's have a try!

3. 模特基础训练 / 模特基礎訓練
Basic Model Exercise

(1) 站姿—五点式贴墙 / 站姿—五點式貼牆
Stance—Stand with five parts of the body clinging to the wall

要做到 / 要做到
Key Points

头、肩、臀、小腿、脚跟贴墙站稳。

頭、肩、臀、小腿、腳跟貼牆站穩。

Your head, shoulders, hips, lower legs and heels should be close to the wall.

萤火虫 / 螢火蟲
Dandelion

让我们一起来玩纸片人的游戏吧！

讓我們一起來玩紙片人的遊戲吧！

Let's play the Paper Man game!

(2) 平移步 / 平移步
Translation Walk

小组合 / 小組合
Small Motion Combinations

准备：双手叉腰，两脚并拢。

準備：雙手叉腰，兩脚併攏。

Get ready: stand with your feet together and hands on the waist.

[1] 1-4 自然抬起左小腿朝1点方向提出，重心完全落在右脚。

[1] 1-4 自然抬起左小腿朝1點方向提出，重心完全落在右脚。

[1] 1-4 Lift your left foot naturally and kick toward Point 1. Put your weight completely on the right foot.

[1] 5-6 左脚落地同时身体保持平稳重心迅速转移至左脚。

[1] 5-6 左脚落地同時身體保持平穩重心迅速轉移至左脚。

[1] 5-6 Shift your weight quickly to the left foot as it reaches the floor and keep yourself balanced.

[1] 7-8 保持动作，身体向上拉伸。

[1] 7-8 保持動作，身體向上拉伸。

[1] 7-8 Remain in the former position, and extend your body upward.

[2] 1-4 自然抬起右小腿朝1点方向踢出，重心完全落在左脚。

[2] 1-4 自然抬起右小腿朝1點方向踢出，重心完全落在左脚。

[2] 1-4 Lift your right foot naturally and kick toward Point 1. Put your weight completely on the left foot.

[2] 5-6 右脚落地同时身体保持平稳重心迅速转移至右脚。

[2] 5-6 右腳落地同時身體保持平穩重心迅速轉移至右腳。

[2] 5-6 Shift your weight quickly to the right foot as it reaches the floor and keep yourself balanced.

[2] 7-8 保持动作，身体向上拉伸。

[2] 7-8 保持動作，身體向上拉伸。

[2] 7-8 Remain in the former position, and extend your body upward.

重复 [1] - [2]。

重復 [1] - [2]。

Repeat [1] - [2].

我能行 / 我能行
I Can Do It

(3) U 型三步转身 / U 型三步轉身
U Turn in Three Steps

要做到 / 要做到
Key Points

双脚并拢，左脚向 1 点迈一步，重心转移至左脚。右脚脚尖朝 7 点迈一步，重心转移至右脚。左脚脚尖朝 5 点迈步，双脚并拢。

雙腳併攏，左腳向 1 點邁一步，重心轉移至左腳。右腳腳尖朝 7 點邁一步，重心轉移至右腳。左腳腳尖朝 5 點邁步，雙腳併攏。

Stand with your feet together. Move your left foot toward Point 1, and shift weight to the left foot. Move your right foot toward Point 7, and shift weight to the right foot. Move your left foot toward Point 5, and stand with your feet together.

我能行 / 我能行
I Can Do It

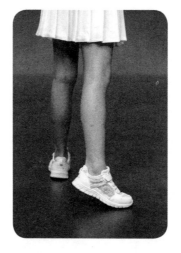

4. 主题秀 / 主題秀
Theme Show

我能行 / 我能行
I Can Do It

直线行走＋u型三步转身

直綫行走＋u型三步轉身

Walk straight forward + make a U turn in three steps

星星班长的话
星星班長的話
Xingxing's Words

第五单元

哈喽，亲爱的孩子们，新的学期开始啦。

过了一个假期，是不是觉得身体的细胞都蠢蠢欲动了呢？

所以在这个单元，训练内容以恢复状态为主，

结合主题秀，让孩子们初步感受走秀的魅力。

第五單元

哈囉,親愛的孩子們,新的學期開始啦。

過了一個假期,是不是覺得身體的細胞都蠢蠢欲動了呢?

所以在這個單元,訓練內容以恢復狀態為主,

結合主題秀,讓孩子們初步感受走秀的魅力。

Unit Five

Hi, dear kids. It's a new semester now.
After a vacation,
are you looking forward to making a move now?
As a result, this unit mainly helps you recover.
With the Theme Show,
kids would feel the charm of walking on the catwalk.

1. 热身训练 / 熱身訓練
Warm-up Exercise

要做到 / 要做到
Key Points

根据音乐节奏完成动作。

根據音樂節奏完成動作。

Complete your body movement according to the rhythm of the music.

与搭档协调配合。

與搭檔協調配合。

Coordinate with your partner to complete the body movement.

小提示 / 小提示
Notes

这样的训练能够提高孩子的学习兴趣,增加协作意识,为后期的搭档走秀做准备。

這樣的訓練能夠提高孩子的學習興趣,增加協作意識,爲後期的搭檔走秀做準備。

This kind of training will rise kids' learning interests, strengthen their sense of coordination, which is an important preparation for their performance on catwalk with partners later.

2. 方位感知训练 / 方位感知訓練
Direction Perception Exercise

(1) 肢体方向练习 / 肢體方向練習
Limb Direction Training

要做到 / 要做到
Key Points

对上、下、左、右、斜上、斜下等空间方位的理解和把控。

對上、下、左、右、斜上、斜下等空間方位的理解和把控。

Be able to understand directions such as the upper, the lower, the left, the right, the left or right upper, and the left or right lower.

小组合 / 小組合
Small Motion Combinations

"Lemone Tree"

(2) 平衡练习 / 平衡練習
Balance Training

要做到 / 要做到
Key Points

腰腹收紧，脚踝与腿部有力量，身体保持拉伸状态，具有稳定性。

腰腹收緊，腳踝與腿部有力量，身體保持拉伸狀態，具有穩定性。

Tighten your waist and abdomen. Make sure that your ankles and legs have much strength. Keep your body extended and balanced.

我能行 / 我能行
I Can Do It

萤火虫 / 螢火蟲
Firefly

小朋友们，快去与你们的爸爸妈妈比一比，谁的平稳性更好！

小朋友們，快去與你們的爸爸媽媽比一比，誰的平穩性更好！

Dear kids, compete with your parents and see who is better at balancing yourself.

蒲公英 / 蒲公英
Dandelion

模特在走秀中，脚踝有力量是非常重要的，这样才能让人感觉走路带风哦！

模特在走秀中，脚踝有力量是非常重要的，這樣才能讓人感覺走路帶風哦！

While walking on the catwalk, the strength of models' ankles is very important, because it'll let audience feel they are walking naturally and quickly.

(3) 出场定位 / 出場定位
Position on the Stage

要做到 / 要做到
Key Points

了解出场顺势定位与逆势定位的区别。

了解出場順勢定位與逆勢定位的區別。

Understand the differences between 顺势定位 **and** 逆势定位.

顺势定位姿势 1　　顺势定位姿势 2　　逆势定位姿势 1　　逆势定位姿势 2　　逆势定位姿势 3

3. 主题秀 / 主題秀
Theme Show

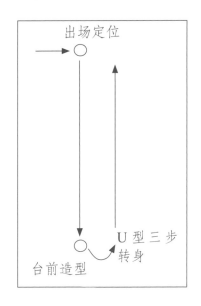

出场定位 + 台前造型 +u 型三步转身 /
出場定位 + 台前造型 +u 型三步轉身
Position on the stage+Pose on the stage + Make a U turn in three steps

要做到 / 要做到
Key Points

熟练掌握走秀路线并具有舞台表现力。
熟練掌握走秀路線并具有舞台表現力。
know your walking route clearly and have stage expressiveness.

我能行 / 我能行
I Can Do It

星星班长的话
星星班長的話
Xingxing's Words

第六单元

孩子们,经过前几单元的学习,
你们是不是已经掌握一些走秀技能了呢?
其实每一场华丽的大秀都离不开舞美灯光,
音乐特效,服饰道具等等的烘托,
更离不开模特在台上精心设计的每一个造型,每一个转身。
让我们学习更多的技能,加油吧!

第 六 單 元

孩子們，經過前幾單元的學習，

你們是不是已經掌握一些走秀技能了呢？

其實每一場華麗的大秀都離不開舞美燈光，

音樂特效，服飾道具等等的烘托，

更離不開模特在台上精心設計的每一個造型，每一個轉身。

讓我們學習更多的技能，加油吧！

Unit Six

Dear kids, have you acquired some skills after
learning the last several units?
In fact, every gorgeous show can not do without stage art,
lights, musical and other effects, clothes and props.
Of course, every pose and turn designed carefully on the
stage by models are also extremely important to the show.
Let's work hard to learn more skills!

1. 热身训练 / 熱身訓練
Warm-up Exercise

要做到 / 要做到
Key Points

熟练掌握跑跳步，后踢步，追步等快速步伐。

熟練掌握跑跳步，後踢步，追步等快速步伐。

Be good at different fast steps such as running and jumping step, kicking step and chasing step.

2. 方位感知训练 / 方位感知訓練
Direction Perception Exercise

要做到 / 要做到
Key Points

动作与音乐相结合，能够跟着节奏进行律动。

動作與音樂相結合，能够跟着節奏進行律動。

Combine body movements with the music, and be able to move according to the rhythm.

间奏的时候可以创造不同的 pose，为造型能力做积累。

間奏的時候可以創造不同的 pose，爲造型能力做積累。

Create different poses during the interludes in order to strengthen the posing capability.

3. 模特基础训练 / 模特基礎訓練
Basic Model Exercise

(1) 三步直线转身 / 三步直綫轉身
Straight Turn in Three Steps

要做到 / 要做到
Key Points

双脚并拢，左脚脚尖朝一点迈步，重心转移至左脚。

雙脚併攏，左脚脚尖朝一點邁步，重心轉移至左脚。

Stand with your feet together. Move the left foot toward Point 1 and shift weight to the left foot.

右脚脚尖朝 7 点，并迈向距左脚脚尖一掌距离。

右脚脚尖朝 7 點，並邁向距左脚脚尖一掌距離。

Point the right foot toward Point 7, and move the right foot to the left one. Make sure the right foot is a little distance away from the left one.

左脚抬起，并将脚尖朝 5 点迈出一小步，重心转移左脚。

左脚抬起，並將脚尖朝 5 點邁出一小步，重心轉移左脚。

Lift the left foot and move it toward Point 1. Shift weight to the left foot.

右脚与左脚并步

右脚與左脚併步

Move the right foot to the left one and stand with your feet together.

   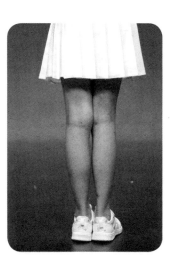

(2) U 型五步转身 / U 型五步轉身
U Turn in Five Steps

要做到 / 要做到
Key Points

双脚并拢，左脚向 1 点迈一步，重心转移至左脚。

雙腳併攏，左腳向 1 點邁一步，重心轉移至左腳。

Stand with your feet together. Move the left foot toward Point 1 and shift weight to the left foot.

右脚脚尖向 7 点迈一步，重心转移至右脚。

右腳腳尖向 7 點邁一步，重心轉移至右腳。

Move the right foot toward Point 7 and shift weight to the right foot.

左脚脚尖朝 7 点迈一步。

左腳腳尖朝 7 點邁一步。

Move the left foot toward Point 7.

右脚继续朝 5 点迈步。

右腳繼續朝 5 點邁步。

Move the right foot toward Point 5.

左脚脚尖继续朝 5 点迈步与右脚并步。

左腳腳尖繼續朝 5 點邁步與右腳併步。

Move the left foot toward Point 5 and stand with two feet together.

4. 主题秀 / 主題秀
Theme Show

出场定位—台前造型—三步直线转身—下场

出場定位—台前造型—三步直綫轉身—下場

Position on the stage—pose on the stage—make a straight turn in three steps—go off stage

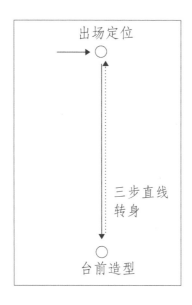

出场定位—台前造型—u型五步转身—下场

出場定位—台前造型—u型五步轉身—下場

Position on the stage—pose on the stage—make a U turn in five steps—go off stage

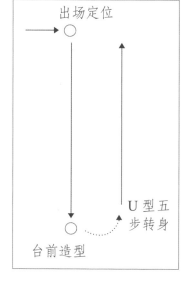

我能行 / 我能行
I Can Do It

第七单元

本单元让我们将学过的内容融合并运用，自信地完成每一个主题秀。下一个超模就是你！

第七單元

本單元讓我們將學過的內容融合並運用，

自信地完成每一個主題秀。

下一個超模就是你！

Unit Seven

In this unit,
we are going to integrate what we've learned and
put them into practical use so as to complete every
theme show confidently.
You, are the next supermodel!

1. 热身训练 / 熱身訓練
Warm-up exercise

要做到 / 要做到
Key Points

完成主题动作"转头""开肩""扩胸""摆胯""伸展""跳跃"等。

完成主題動作"轉頭""開肩""擴胸""擺胯""伸展""跳躍"等。

Complete theme movements such as turning the head, extending the chest, twisting the hip, stretching and jumping.

我能行 / 我能行
I Can Do It

《蜡笔小新》

《蠟筆小新》

"Crayon Shin-chan"

2. 方位感知训练 / 方位感知訓練
Direction Perception Exercise

要做到 / 要做到
Key Points

听到音乐的同时，通过想象和模仿用肢体表现出来。

聽到音樂的同時，通過想象和模仿用肢體表現出來。

When you hear the music, you should be able to express the music with your imagination and imitation.

小提示 / 小提示
Notes

强大的表现力是模特必备条件之一，自信心和表现力息息相关，感知训练的作用在后期的学习中尤为重要，可以大大提升表演的自信心。

强大的表現力是模特必備條件之一，自信心和表現力息息相關，感知訓練的作用在後期的學習中尤為重要，可以大大提升表演的自信心。

Strong expressiveness is essential to a model, while confidence is closely related to expressiveness. As a result, the perception exercise is especially important in the learning of the later stage, for it can greatly enhance one's confidence while performing.

3. 模特基础训练 / 模特基礎訓練
Basic Model Exercise

(1) 双直角转身训练 / 雙直角轉身訓練
Double Right Angle Turn Training

要做到 / 要做到
Key Points

在每个角度转身的时候自然，不生硬。

在每個角度轉身的時候自然，不生硬。

Turn around naturally at every angle.

(2) 后台两步定位 / 後台兩步定位
Position in Two Steps at the Back of the Stage

要做到 / 要做到
Key Points

左脚向 5 点迈步，右脚向 3 点迈步且脚尖朝 1 点方向，同时左脚脚尖转向 1 点，右脚平行于左脚，并将重心转移到右脚，左脚脚尖转向 8 点。

左腳向 5 點邁步，右腳向 3 點邁步且腳尖朝 1 點方向，同時左腳腳尖轉向 1 點，右腳平行於左腳，並將重心轉移到右腳，左腳腳尖轉向 8 點。

Move the left foot toward Point 5. Move the right foot toward Point 3 and point it toward Point 1. In the meantime, point the left foot toward Point 1, so as to make two feet parallel. Shift weight to the right foot, and point the left foot toward Point 8.

右脚向 5 点迈步，左脚向 7 点迈步且脚尖朝 1 点方向，同时右脚脚尖转向 1 点，左脚平行于右脚，并将重心转移到左脚，左脚脚尖转向 2 点。

右腳向 5 點邁步，左腳向 7 點邁步且腳尖朝 1 點方向，同時右腳腳尖轉向 1 點，左腳平行於右腳，並將重心轉移到左腳，左腳腳尖轉向 2 點。

Move the right foot toward Point 5. Move the left foot toward Point 7 and point it toward Point 1. In the meantime, point the right foot toward Point 1, so as to make two feet parallel. Shift weight to the left foot, and point the left foot toward Point 2.

4. 主题秀 / 主题秀
Theme Show

要做到 / 要做到
Key Points

出场定位—台前造型—u 型三步转身—后台两步定位

出場定位—台前造型—u 型三步轉身—後台兩步定位

Position on the stage— pose on the stage— make a U turn in three steps— position in two steps at the back of the stage.

我能行 / 我能行
I Can Do It

星星班长的话
星星班長的話
Xingxing's Words

第八单元

中国是"礼仪之邦",
模特与礼仪也是密不可分的,
日常礼仪常识是每个积极向上的少年儿童应具备的基本素养。
本单元着重介绍个人礼仪常识,
让我们成为既有良好个人修养又能进行模特艺术表演的小超模吧!

第八單元

中國是"禮儀之邦",

模特與禮儀也是密不可分的,

日常禮儀常識是每個積極向上的少年兒童應具備的基本素養。

本單元著重介紹個人禮儀常識,

讓我們成為既有良好個人修養又能進行模特藝術表演的小超模吧!

Unit Eight

China is a state of etiquette.
Etiquette is also inseparable to models.
Following daily etiquette should be the
basic quality of every positive children.
This unit mainly introduces some common
sense of personal etiquette.
Let's become supermodels who not only perform
professionally but also have good personal etiquette.

1. 模特展示技巧 / 模特展示技巧
Demonstration Skills for Models

女生常用 pose

女生常用 pose

Commonly-used poses for girls

模特初级教程 / 模特初级教程
Tutorial for Model Beginner

男生常用 pose

男生常用 pose

Commonly-used poses for boys

模特初级教程 / 模特初级教程
Tutorial for Model Beginner

2. 日常礼仪 / 日常禮儀
Daily Etiquette

(1) 微笑 / 微笑
Smile

要做到 / 要做到
Key Points

面部肌肉放松，嘴角微微上翘，不发出笑声，不露出牙齿和牙龈，目光有神，真诚地展现笑容。

面部肌肉放鬆，嘴角微微上翹，不發出笑聲，不露出牙齒和牙齦，目光有神，真誠地展現笑容。

Relax facial muscles, smile slightly without making sound or showing teeth and gums. Make sure to have bright eyes and show sincere smiles.

我能行 / 我能行
I Can Do It

蒲公英 / 蒲公英
Dandelion

想象美好的场景,眼神里自然会流露快乐的情感。微笑是最好的名片,谁都希望能交到一个乐观、积极向上的朋友,让我们一起微笑吧!

想象美好的場景,眼神裏自然會流露快樂的情感。微笑是最好的名片,誰都希望能交到一個樂觀、積極向上的朋友,讓我們一起微笑吧!

Imagine a happy scene, so there will be a sense of happiness in your eyes. Smile is best way to express yourself. All want to make friends with optimistic and positive people. Let's smile!

(2) 交谈 / 交談
Communication

要做到 / 要做到
Key Points

轻声细语，目视对方，注意倾听，使用文明用语。

輕聲細語，目視對方，注意傾聽，使用文明用語。

Talk in a low voice. Have eye contacts with the one your are talking with. Be a careful listener. Use polite words.

男生坐姿

女生坐姿

(3) 坐姿 / 坐姿
Sitting Posture

要做到 / 要做到
Key Points

双膝并拢，小腿不可一前一后，双脚平放落地不要交叉晃动。

雙膝併攏，小腿不可一前一後，雙腳平放落地不要交叉晃動。

Sit with your knees and legs together. Do not cross or move your feet.

上身与桌椅背保持 10-15 厘米距离，双手掌心向下放于膝盖。

上身與桌椅背保持 10-15 公分距離，雙手掌心向下放於膝蓋。

Keep the upper body 10-15cm apart from the table and chair. Place your hands on the knees with palms down.

坐姿交谈时，上身与双腿应同时转向对方，目光正视对方。

坐姿交談時，上身與雙腿應同時轉向對方，目光正視對方。

If you are talking to others while sitting, your should turn your upper body and legs toward the one your are talking with and have eye contact with him or her.

Postscript

爱美是每个人的天性，孩子也不例外，而模特在T台和生活中更是美的承载者与模范者。我希望孩子们通过系统的课程学习不仅都能够由内而外，在耀眼的舞台上，秀出自信与美丽，并且能将这份专业素养与时尚气质带入日后的人生舞台中。

愛美是每個人的天性，孩子也不例外，而模特在T台和生活中更是美的承載者與模範者。我希望孩子們通過系統的課程學習不僅都能夠由內而外，在耀眼的舞台上，秀出自信與美麗，並且能將這份專業素養與時尚氣質帶入日後的人生舞台中。

The desire of beauty is human nature, and children are no exception. No matter on the catwalk or in real life, models demonstrate beauty. I hope that through a series of courses, kids can not only display their confidence and beauty on gorgeous stage, but also maintain the professional quality and fashion temperament in their lives.

模特在教学工作中能用形象、科学、系统的文字来表达是任重而道远的。在此我要感谢小荧星教研组全体一线教师给予我的许多积极建议，感谢小荧星艺术学校的小学员们为本教程的图片和视频内容做示范表演，感谢上海科学技术文献出版社对本书出版给予的支持和帮助。

模特在教學工作中能用形象、科學、系統的文字來表達是任重而道遠的。在此我要感謝小熒星教研組全體一線教師給予我的許多積極建議，感謝小熒星藝術學校的小學員們爲本教程的圖片和視頻內容做示範表演，感謝上海科學技術文獻出版社對本書出版給予的支持和幫助。

We have a long way to go to teach modeling with vivid, scientific and systematic words. I would like to extend my heartfelt gratitude to all the teacher of Shanghai Little Star for giving me so many suggestions, to all the students for participating in the making of pictures and videos, and to the Shanghai Science and Technology Publishing House for supporting the publication of this book.

本书是"小荧星艺术学校通用教程"的初级教程,之后还会有中级、高级教程陆续出版。

本書是"小熒星藝術學校通用教程"的初級教程,之後還會有中級、高級教程陸續出版。

This book is the elementary textbook of the general textbooks of Shanghai Little Star Art School. The intermediate and advanced textbooks will be published later.

水 芳

水 芳

Shui Fang

2018.6

星 星 墙 / 星 星 牆
Wall of Little Stars

第一单元 / 第一單元
Unit One

第二单元 / 第二單元
Unit Two

第三单元 / 第三單元
Unit Three

第四单元 / 第四單元
Unit Four

第五单元 / 第五單元
Unit Five

第六单元 / 第六單元
Unit Six

第七单元 / 第七單元
Unit Seven

第八单元 / 第八單元
Unit Eight